Integrate Your Brain

The Basics of Doing Integral Calculus In Your Head

Aaron Maxwell

Published by Hilomath (http://hilomath.com).

This book was set in the Times font using LaTeX, and written with the LyX editor (http://lyx.org). The figures and cover were assembled using The GIMP (http://gimp.org). The math symbols in the figures were created with the L2P tool (http://red-symbol.net/software/l2p/).

We like to hear from you. For publisher correspondence, please send email to service@hilomath.com, or post mail to Hilomath, Box 225120, San Francisco, CA 94122, USA.

Contents

Preface

This book is mainly written for someone who has completed a first calculus course, and assumes that you know how to symbolically integrate expressions of one variable. If you are taking your first calculus course right now, however, you can still use this book as a companion. I recommend you only use the techniques in this book to do math that you have already mastered in your coursework. This is to help make sure you learn the math well: when working some math problem that is new to you, writing out all the steps is the best way to cement your understanding and ingrain the correct habits. Once you have that clarity, you have a great foundation on which to do that kind of abstract math mentally. If you have any questions about this process, please show this book to your instructor and ask them for guidance.

Several places in this book refer to another book, called *Inner Algebra,* which teaches how to do algebra mentally. If you do not happen to have it, you can read the full text online at http://hilomath.com/inneralgebra/html/ .

While the aim of this book is to make you skillful at doing abstract math in your head, please know that it is not an all-or-nothing affair. At any time when doing mental math, you can always choose to write down a step to make sure you are clear. The most important thing when doing math is to maintain your understanding - never forget that!

If you have questions or comments, I'm happy to hear from you. Contact me by sending email to amax@hilomath.com.

Acknowledgements

Thanks to the readers of my previous book, *Inner Algebra*, for valuable feedback. Thanks to my friend Sarah Hwang for helping with the title and cover. (The results are much better than my original ideas!)

I'd like to express my gratitude to the worldwide community of free and open source software developers. This book was written and typeset using open source software, including the L$_Y$X editor (along with the memoir document class), the L^{A}T$_E$X typesetting system, the GIMP image editor, and the Debian GNU/Linux operating system (with all of its supporting software). This is the fruit of the cumulative, generous efforts of a large number of software developers, who over many years donated their time to help create the software used every day by myself and many others. Thank you.

Chapter 1

Introduction: Mental Foundations

This short book teaches how to evaluate many integrals mentally. For the most part, if you can now evaluate an integral by writing it out on paper, you can potentially learn to solve it just through thought. With relatively complex integrations, in which you would normally need to write out many steps to solve, you will be able to solve by writing just a few steps down.

You won't learn here to solve every integral that exists; many real-world integrals can only be evaluated through software, and the methods here will not solve those. Fortunately, what you learn will apply in some way to essentially every integral you come across. Even if it is the kind that must be evaluated with a computer, these methods will help you work with them more easily, with greater intuition and understanding.

When you are learning to solve some math that is new to you, there is a process that you go through. At first, you will solve problems by writing out several steps. As you get better at doing that particular kind of math, you can go through other, similar examples of that kind more quickly and easily. As part of this, you tend to write out fewer intermediate steps in the process of solving it.

Some people find it very easy to do more and more of the steps in their head. Because of how their minds work, abstract math comes naturally to them. They also find relatively arcane math more accessible. That's not to say it isn't hard work for them - generally, it is! But some of the things they do inside serve to make the simple math easier, and allow the harder things to be *possible*.

These characteristics are based in common mental abilities that nearly every human has. In the following chapters, you will learn about some of these characteristics, and how to develop them in yourself.

Given an integral, to evaluate it you want to rearrange it into a form $\int f(u) \, du$, where f is some function whose antiderivative $F(u)$ is known to you. That's the abstract; concrete examples of this would be $f(u) = \sin u$ and $F(u) = -\cos u$, or $f(u) = 1/u$ and $F(u) = \ln u$, or $f(u) = u^2 + 2u^3$ and $F(u) = \frac{1}{3}u^3 + \frac{1}{2}u^4$.

When you learn to do integration in a calculus course, one of the first things you are taught is to manage complexity by doing variable substitution. Say you were integrating $\int \sin x^2 \, 2x \, dx$; by inspecting the equation, you might think to try setting $u = x^2$, notice that $du = 2x \, dx$, and rewrite the expression to be $\int \sin u \, du$. After evaluating it to $-\cos u + C$ (where C is a constant), you would plug back in to get $-\cos x^2 + C$.

All the steps described above can be done easily enough on paper. We can do something quite similar mentally, in a way that scales to more complex integrands. A big key is to manage it in a way that you can visualize the expression as well as the processes of evaluating it.

If your visual imagination is very highly developed, you can do complex math mentally just as you would when solving it on paper. You could just imagine a sheet of paper (or whiteboard, chalkboard, etc.), and imagine all the symbols that you would write on the paper physically as you evaluated the expression. However, many people would have trouble doing this with larger expressions (or even with small ones). Even if you are able to visualize that well, there are better ways. A piece of paper is one medium; your imagination is a different one, and one that is more flexible. If you envision a whiteboard in your mind, with mathematical symbols written on it, you can do things with this mental representation that would be impractical or impossible if you were writing on a physical whiteboard. In fact, some of these things you can do are very powerful, and can more than make up for the fact that you don't have a written record in front of you.

As a human, there are a few mental resources you have that you can use. The first is your ability to visualize. Some people seem to be born with an ability to visualize well; others don't seem to consciously visualize at all. Most people are somewhere in between. If you do not feel very strong in this area, there is a visualization tutorial available at http://hilomath.com/inneralgebra/html/node7.html . (This tutorial is part of *Inner Algebra*, a book that teaches how to solve algebra equations mentally.)

Another mental resource you have is your memory. You can use your memory in a few specific ways that will help you manage the mental process you go through. A key one is called *windowing*. Rather than visualize the entire expression or equation, you visualize a portion of it that you want to work with. Later, you can visualize the rest of it or another part of it, recalling it from memory as you need to. This works because when working with an expression, you often only need to alter a few of the symbols; most of the expression will not be affected, at least for that one step. Windowing is based on the fact that most people can recall a math expression that is bigger than the largest one they can easily visualize. You can also do this with smaller expressions that you CAN easily visualize. In fact, it's highly recommended. If the expression is small enough that you can visualize the whole thing, you choose to visualize only part of it at any given moment. Then, when you need to work with another part of the expression, or the whole thing, you can consciously construct the image from your memory. Simple as it may sound, this can be tremendously helpful in practice, allowing you to do mental math much faster. (You can read a more in-depth discussion on windowing, also from *Inner Algebra*, at http://hilomath.com/-inneralgebra/html/node23.html . That section also has some short but valuable exercises.)

Chapter 2

Chunks and Chunking

There is a kind of mental device you can use called a *chunk*. We also speak of *chunking*, which is the action of creating a chunk. It's basically a tool for organizing your attention. You make a chunk by taking a group of symbols in the expression you visualize, and work with that group as a logical unit. Usually you would "decorate" it in some way, meaning that you would visualize that portion of the image a little differently. For example, you could visualize the chunk (group of symbols) as having a dotted line enclosing it, or visualize those symbols on a differently colored background from the rest of the expression.

When you chunk a group of symbols - and again, chunking is a *mental* action - you are basically saying that you intend to do something mathematically with them. Suppose you want to evaluate the integral $2a \int \sin ax^2 \, x \, dx$, where a is a constant. To do it mentally, you would visualize the expression:

$$2a \int \sin ax^2 \; x \; dx$$

(In this book, images you visualize internally are shown in a box like this.) Looking at it, you notice that the derivative of ax^2 is $2ax$. So one of the next steps is to determine if that derivative is part of the integrand, and if so, to group the differential $2ax\,dx$ together. To manage this mentally, you would chunk the 2a on the left:

$$2a \int \sin ax^2 \; x \; dx$$

(We denote a chunk by putting a dotted box around it. You can visualize it this way yourself internally. As stated above, though, there are other valid ways to do it. You don't need to visualize it exactly as shown in the figure, if you find a way that seems to work better for you.)

Now that you have the chunk, you can move it to a better location in the expression:

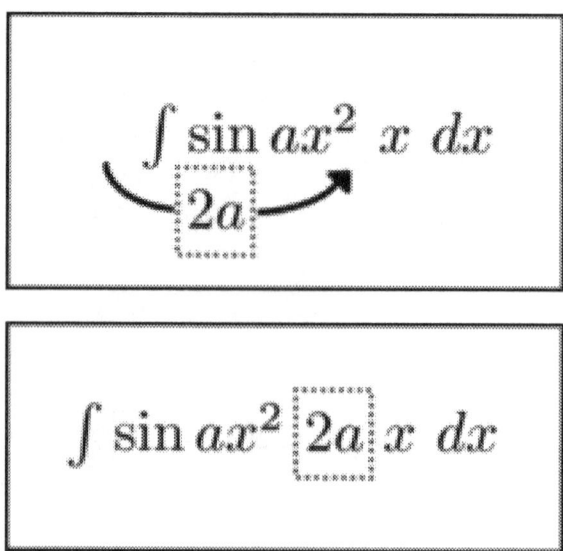

What you see in your mind's eye is the 2a chunk actually moving from the spot on the left, sliding into place just before the $x\,dx$ (and scooting the $x\,dx$ to the side in the process). Now, when you were first learning to do calculus, you would have written a line that looks like

$$2a \int \sin ax^2\, x\, dx$$

and then, underneath it, written

$$\int \sin ax^2\, 2ax\, dx$$

Moving the chunk mentally accomplishes the same thing... which is, of course, the whole point.

When integrating expressions, we sometimes use variable substitution to manage the complexity. Let $u = ax^2$, and $du = 2ax\,dx$; the integral then becomes the simpler expression $\int \sin u\, du$. Something similar can be done when evaluating the expression mentally, using chunks. The first step is to decide on the variable change - in this case, $u = ax^2$. In the expression, chunk the set of symbols that will be replaced by the new variable:

$$\int \sin ax^2\, 2a\, x\, dx$$

Then visualize the chunked part being replaced with a new chunk, containing just the variable:

$$\int \sin u\, 2ax\, dx$$

   Swapping chunks like this is called *chunk substitution*. To complete the process, we must also replace the differential. Chunk the symbols corresponding to that,

$$\int \sin u \; \boxed{2ax \; dx}$$

then substitute the chunk.

$$\int \sin u \; \boxed{du}$$

   At this point it's simple to evaluate the expression and change back to the original variable:

$$-\cos \boxed{u} + C$$

$$-\cos \boxed{ax^2} + C$$

(C is the constant of integration, of course.)

There are many other things that can be done with chunks, such as cloning or copying them, moving them within the expression in different ways, creating several chunks in one expression (decorated differently so as to distinguish them), and more. It's a great organizational tool for doing higher mental math.

The best thing about chunks, though, is that as you use them, they train you to not need them. Chunks are to math much like training wheels are to riding a bike. As you do things with chunks many times, gaining confidence and experience, you will learn to do the kinds of mathematical transformations they produce more intuitively, without having to explicitly use chunks.

Let's look again at our example of the last section, $2a \int \sin ax^2 \, x \, dx$. Jump to the step where we have rearranged the symbols, and are now ready to do the variable substitution. You are visualizing something like this:

$$\int \sin \boxed{ax^2} \, 2a \, x \, dx$$

We have already chunked the argument of the sine function here. This time, instead of doing a variable substitution, we will just work with chunks. Since $u = ax^2$ and $du = 2a \, x \, du$, you can also write du as $d(ax^2)$. Visualizing the expression to reflect this, it looks like

$$\int \sin ax^2 \, d\left(ax^2\right)$$

(Writing differentials like this is common in advanced math texts, by the way.) You can then chunk both occurrences of ax^2:

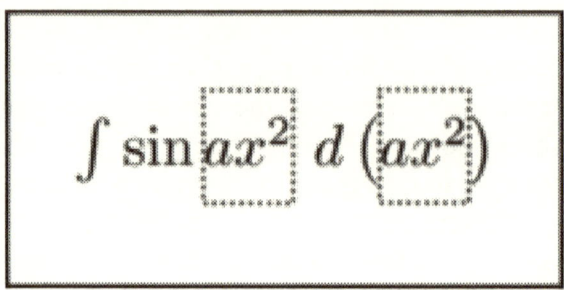

Now, if you wanted to, you could do chunk substitution with both, replacing ax^2 with u. But you don't have to here. Variable substitution is a simplification, which you introduce to make the expression more manageable. When you chunk like this, though, you have already done something like that. In fact, at this point, you can visualize the expression like this:

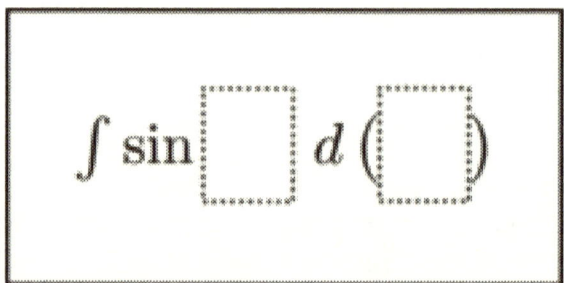

See what we did? Semantically, this is just like visualizing $\int \sin u\, du$. In case you are not familiar with the word, "semantics" refers to the *meaning* attached to symbols in a printed or visual form[1]. ax^2 and the symbol u have the same mathematical meaning, so we say they are semantically equivalent. In this expression, the empty chunk represents ax^2, and has the same

---

[1]The word "semantics" actually has a broader definition than this. For our purposes, the definition we use is fine.

mathematical meaning too. Thus, the empty chunk is semantically the same as the expression ax^2.

Incidentally, what we are doing above is a special kind of windowing. We are visualising only that part of the expression which we need to at this time. We really don't need to visualize the ax^2, so we don't. We know the empty chunk represents it.

When you integrate, you can visualize this:

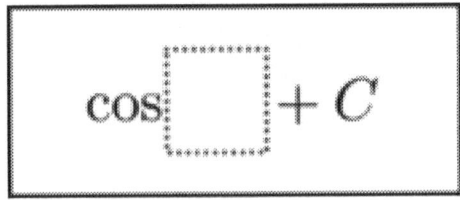

Even though you have transformed the expression, you are still windowing, because you are still excluding the ax^2 from the image. At this point, you can recall the chunk's contents.

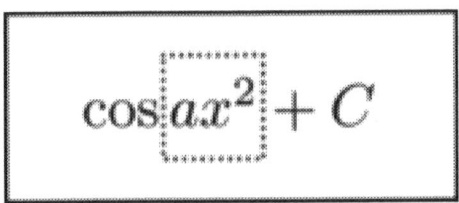

The empty chunk is not quite the same thing as a regular chunk; it is a slightly different animal, acting in some respects like a placeholder. Remember, chunks are basically a device to help you *organize* and *direct* your *attention*. Most people think of attention as just concentrating on one thing or another. Attention also has a more subtle meaning, that includes the *patterns* and *habits* of exactly how you are directing that concentration. Using chunks trains you to use your attention in certain precise ways - ways that happen to be used by good mathematicians, engineers, and hard scientists.

Again, chunks are like training wheels on a bike. When you first get on a bike, you might need the training wheels in order to ride at all. They're great; you can ride really fast on this bike because of them. As you ride more, your skill and awareness grow, and you rely on the training wheels less and less. When you get good enough to do really fancy things, you might even find that they get in the way, so you don't use them. If you ever need them again in the future, though, they are always available.

The abilities you have between your ears are more expressive than can be put into words. What is written in these pages is meant to point you in the right direction, so that you can discover the rest for yourself as you do it.

Chapter 3

Integration by Parts, In Your Head

This section explains how to mentally do integrations which must be solved by integration by parts. As you read, it may seem that you have to keep an unreasonable amount of information in your head. "How am I supposed to keep track of all this? It's not possible." What you are learning to do is work with your *internal representation* of the math. If you visualize an equation, that is one possible internal representation. The more efficient the representations you employ inside of you, and the more efficient your methods for working with them, the more complex the math you can easily do. One thing talented mathematicians have working for them is that they have discovered how to internally represent math expressions in ways that are both efficient and amendable to mathematical manipulation. This lets them use the mental energy they have very, very effectively. As mentioned earlier, if a person's ability to visualize is good enough, they can just imagine a whiteboard in their mind's eye, and imagine each step written out exactly as they would write it out on a real whiteboard. That's certainly one possible internal representation, and one that requires a LOT of mental work! Mathematicians learn to represent expressions in more minimal ways, making them much easier to manage. That is precisely what you are learning to do here!

Ease into it gradually. Use the techniques to skip single steps that you would normally write out. For a problem that you might typically write six steps to solve, do one of them mentally and write five steps. After you have done this enough that it seems straightforward, look for opportunities to do two steps in a row mentally. Bite off what you can chew. It's very likely that there will be times when you attempt to do something mentally, and partway through it discover that it was too much. No problem; I do that all the time. So does every mathematician and engineer I know of. Just back off to where you started that step and write it out instead, or break it up into smaller, more manageable pieces.

What's written below is presented in the most extreme case: how you would evaluate the whole expression mentally. You can always use it partially, writing out as many steps as needed. When evaluating a mathematical expression, the most important thing is to understand what you are doing. Do as much or as little mentally as it takes to support that goal.

Most modern calculus textbooks use the symbols u and v when teaching integration by parts. The equation you memorize is probably $\int u \, dv = uv - \int v \, du$, not $\int a \, db = ab - \int b \, da$, even though mathematically they are the same. In addition, the *position* of the symbols is always the same: you typically see and write the right hand side as $uv - \int v \, du$, and not like $vu - \int v \, du$ or $-\int v \, du + uv$.

You can use this consistency to help you do integration by parts mentally. We'll demonstrate with the integral

$$\int \frac{\ln x}{x\sqrt{x}} dx$$

First, you want to rearrange the symbols so that you can work with the expression more clearly. You might end up visualizing something like this:

$$\int x^{-\frac{3}{2}} \ln x \, dx$$

In this form, the expression is organized in a more minimal way, and will be easier to work with.

(Note: while doing algebra mentally is not explicitly covered in this work, it can be done with the techniques covered in the first few sections - visualization, using chunks, and so on. The book *Inner Algebra* (http://hilomath.com/inneralgebra/html/), and the upcoming book *Higher Mental Math,* teach mental algebra in detail.)

When doing integration by parts, you divide the integrand into two parts - one of which will be differentiated, the other integrated. The next step is then to divide up the integrand and differential into two chunks like that:

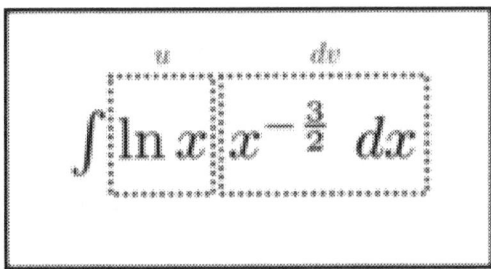

This is what you might visualize. There are a few things to point out. We have mentally moved some of the symbols around. Also, we have decorated the chunks, by putting a little "u" or "dv" on top of them. You can actually visualize these labels over each chunk. Yet you don't have to: the other thing to notice is that we put the u chunk on the left and the dv chunk on the right. Now, as we mentioned earlier, when you learn integration by parts, typically the formula you use has a left-hand side of $\int u\ dv$, rather than $\int dv\ u$. Even though mathematically both versions have the same meaning, you have made a habit of using the first form. (Or if you are taking your first calculus course right now, you will be taught to make it a habit.) You can exploit this habit; all you have to do is divide the expression into two chunks, and put the one corresponding to u on the left of the chunk corresponding to dv. So you can visualize the expression like this:

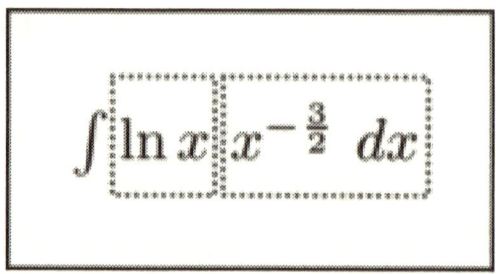

While the chunks are not labelled, you know which is which by its position in the expression. *The position itself acts as the label!* Having fewer symbols, it is easier to visualize and work with mentally. The fewer symbols an expression has, and the simpler they are, the greater the ease with which you can work with that expression.

You can harness this positioning in a powerful, interesting way. Visualize the right side of the integration-by-parts rule:

$$uv\ -\int v\ du$$

Now, replace that with this image:

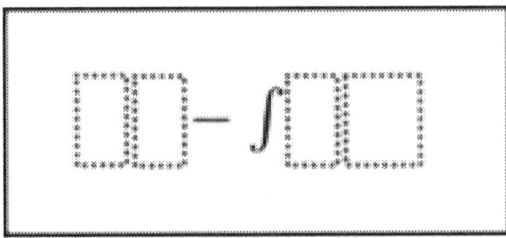

You are visualizing empty chunks, arranged in a pattern like the reference image. Obviously, each of these empty chunks corresponds to u, v, or dv. What you want to do is be able to switch between both images in your mind's eye. Practice this a few times right now.

So now you have two images you are working with: the first one, containing $uv - \int v\,du$, and the second one, containing empty chunks. You mainly work with the second image, filling in the chunks. When you need to remember what goes in each chunk, you visualize the $uv - \int v\,du$ image, so that each symbol (u, v, and dv) overlaps the chunk they correspond to. This can be done in a way that efficiently reminds you of the meaning of each chunk.

Let's examine how this can be done with $\int x^{-3/2} \ln x\,dx$, step by step. Once you have sorted the expression into two chunks, visualize something like this:

$$\int \boxed{\ln x}\,\boxed{x^{-\frac{3}{2}}\,dx} = \boxed{}\boxed{} - \int \boxed{}\,\boxed{}$$

At any time, to remind yourself of the meaning of each chunk, you can replace that image with this one:

$$\int u \, dv = uv - \int v \, du$$

Swap between them in your mind's eyes. This may or may not be easy for you at first. If not, practice!

Since some of the chunks repeat - e.g., u and v both appear twice - you can just copy the chunks to those spaces. Moving left to right, you would first visualize

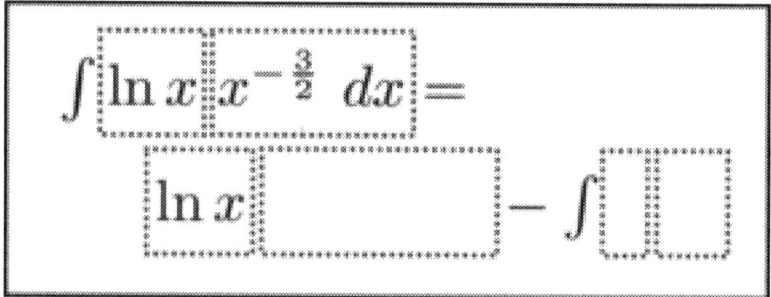

(This image is split up into two lines, so that it can fit on the page you are reading. As you visualize it internally, you can either keep it as one long line or split it up like this. If you do, you may want to split up the image of $\int u \, dv = uv - \int v \, du$ similarly.) We don't have to think much for this first substitution, since we are simply copying the contents of one chunk to another. Schematically, it works like this:

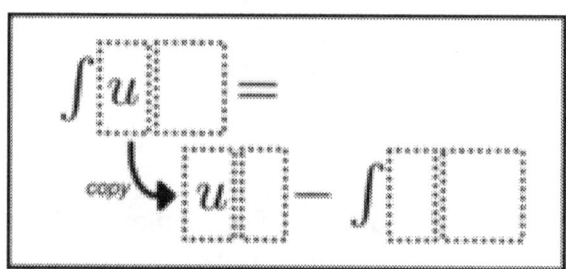

You are taking a part of the mental image you have, and repro-
ducing it in another location. To fill the next chunk on the right
hand side, you will have to integrate the contents of the dv chunk
on the left-hand side. You would end up visualizing something
like this:

$$\int \ln x \; x^{-\frac{3}{2}} \, dx =$$
$$\ln x \, (-2x^{-\frac{1}{2}}) - \int \boxed{}$$

At this point, you can remind yourself of what to do next by
swapping back to this image:

$$\int u \, dv = u v - \int v \, du$$

Since the "v" chunk appears twice on the right hand side, and
we filled the first one, we know what to put in the second one.
Switching back to the other image,

$$\ln x \, (-2x^{-\frac{1}{2}}) - \int (-2x^{-\frac{1}{2}})$$

Note that we only visualize the right hand side here, since for this step that is all we need. Switching again to this image,

$$\int \boxed{u}\,\boxed{dv} = \boxed{u}\,\boxed{v} - \int \boxed{v}\,\boxed{du}$$

we see that the last chunk to fill is the differential of $\ln x$. While mental differentiation is not covered in this book, it is based on all the same principles you have learned so far. (Mental differentiation will be detailed in the upcoming book, *Higher Mental Math*.) Differentiating an expression is often easier than integrating it, fortunately. To finish, copy the u chunk, differentiate its contents, and move it into the last space:

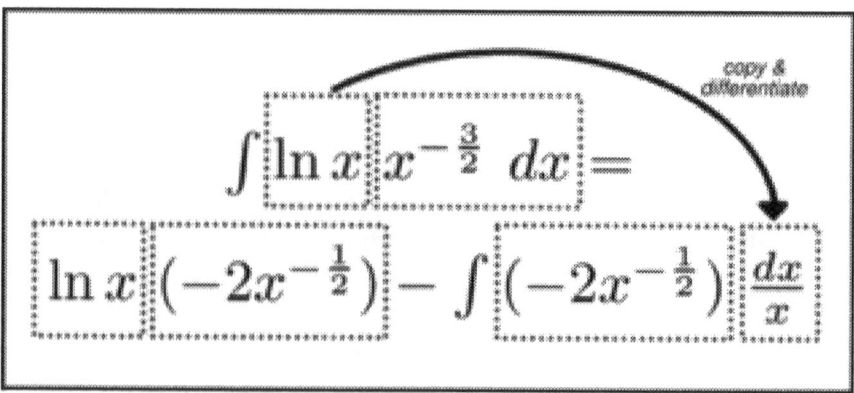

This is a good place to simplify the expression. If you intend to write down the final result, you can simplify and write down the first term on the right-hand side at this point. In fact, you can do it even before now, so that you have less to remember. Let's say you have done this - you have just written something like $-\frac{\ln x}{\sqrt{x}}$, and now you are visualizing this:

$$- \int \left(-2x^{-\frac{1}{2}} \right) \frac{dx}{x}$$

(This is all you have to visualize now, since you have written down the rest and have finished with it.) We still have the chunks there. At this point, though, we no longer need them. So just visualize the expression without them:

$$- \int \left(-2x^{-\frac{1}{2}} \right) \frac{dx}{x}$$

You can simplify it, and perhaps end up visualizing something like this:

$$2 \int x^{-\frac{3}{2}} \, dx$$

To evaluate this mentally, first recognize that this expression fits the rule

$$\int x^n \, dx = \frac{x^{n+1}}{n+1}$$

(plus a constant, but since we are doing integration by parts, we can set this constant to 0.) To mentally integrate a monomial like this[1], begin by visualizing the expression and chunking the exponent:

$$2 \int x^{\boxed{-\frac{3}{2}}}\, dx$$

Then, drop the integral sign and the differential dx from the image:

$$2 \quad x^{\boxed{-\frac{3}{2}}}$$

(By the way, this is what we call an *intermediate image* – an image you visualize that has no (correct) mathematical meaning, but is something you use as a step when moving from one valid image to another.) Now, add one to the exponent (the chunk's contents):

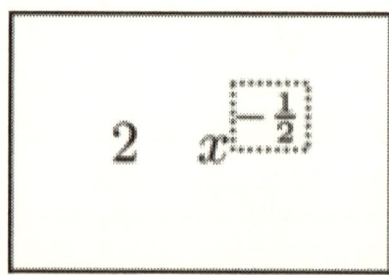

$$2 \quad x^{\boxed{-\frac{1}{2}}}$$

---

[1]By extension, this works with polynomials also.

There is one more step, which can be done in a couple of ways. In the first way, you would make a fraction, copying the chunk and moving it into the denominator:

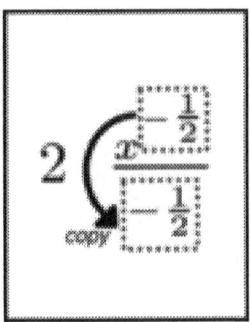

From this point you just de-chunk and simplify:

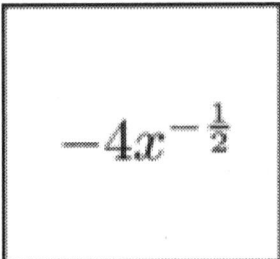

The second way for the last step also involves copying the chunk. But you don't create a fraction; instead, you *invert* the chunk's contents and insert it multiplicatively in the expression:

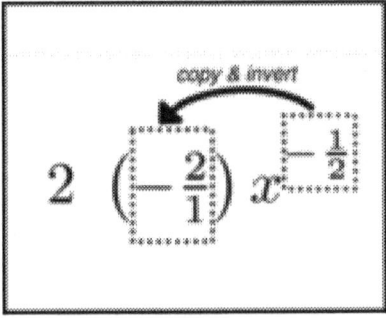

Then, you just de-chunk and simplify. Obviously, these are both mathematically the same. The second way is useful when the exponent is a fraction. The first way is better when the exponent is not a fraction. While you will use both, you will probably use the first way more often.

After all this talk about the internal process, it may be helpful to step back and see what all this looks like from the outside. If you are watching yourself, after you have learned and mastered these techniques, what would you be doing? Imagine you are evaluating an integral that is written on a whiteboard (or written on a sheet of paper, typed into a computer, etc.). You glance at the expression several times as you evaluate it. You may rewrite the expression once with the symbols sorted into two groups, u on the left and dv on the right. Or you may not rewrite it at all, relying on your mental image instead. You write an equal sign, followed by a term that corresponds to uv: it's a simplified expression, with the product of u and v reduced to its simplest form. Next you write an integral, whose argument corresponds to v du. Alternatively, you write the result of evaluating the integral directly, evaluating it internally first.

Chapter 4

Definite Integrals and Number Crunching

If you are evaluating a definite integral, the final step is to evaluate it across its limits of integration. This is basically a matter of arithmetic. This work is mainly concerned with more abstract math; mental arithmetic has already been covered well by others, and there are many good books on the subject. One in particular I like and recommend is *Math Magic* by Scott Flansburg. Nevertheless, there are some techniques and things to consider that are particular to doing abstract math, and which I haven't seen elsewhere. This section discusses some of them.

The best approach depends on the situation. Do you need an exact answer, or will an approximation do? Suppose the correct result is, say, 14.21; if you get 14 as an answer, or even 15, and you know that your answer is close but not exact, will that be good enough for what you need now? The answer really depends on what why you are solving this particular problem. If all you need is an estimate, an approximation will be fine. If you are calculating the final measurements for part of a robotic space satellite, it's important to be as accurate as you can.

If an exact result is necessary, you may be able to calculate it mentally. Again, books such as *Math Magic* teach how to do this even for complex calculations. Otherwise, you can use a calculator or math software for the final, calculation step.

If an approximation is sufficient, good! You are free to take shortcuts that will let you get to the solution quickly. You can round numbers to one or two significant digits (replacing 19.87 with 20, or 1,311 with 1300). You can estimate square roots for non-square arguments; since $\sqrt{19.2}$ is partway between 16 and 25, you might estimate it to be 4.4 or 4.5. (The actual square root is approximately 4.38.)

If you do decide to do the final number calculation mentally, there are two general strategies that make it easier and faster. They are especially useful when you are approximating.

Calculating Term by Term

Here is the first calculating strategy: if the resulting expression has many terms added and/or subtracted together, you can calculate the terms one at a time, keeping a running total as you go along. For example, say you mentally evaluate $\int_2^{10} (u^3 + 2u - 3)\, du$, and end up visualizing

$$\frac{u^4}{4} + u^2 - 3u \Big|_2^{10}$$

It can get complex here, because to evaluate it mentally, there are three groups of information to keep in your head at once:

1. the expression $\frac{u^4}{4} + u^2 - 3u$,

2. the limits of integration, 10 and 2, and

3. the running total.

While it's possible to keep track of all of this information mentally, generally speaking most people will find it hard at first, except for simpler expressions. As you practice more, the size of expression that you consider "simple" will naturally grow. So what do you do when you are dealing with an expression so large that you have trouble remembering everything? Fortunately, you can often rely on a few things that will help. First, many times when evaluating an integral, the original expression will be written or printed someplace that you can see. In our example, $\int_2^{10} (u^3 + 2u - 3)\, du$ will be physically printed somewhere you can easily glance at. This is not always true; sometimes you will have constructed the integral internally yourself. But if it is available physically, you can use it in a few ways. The limits of integration are in it; you can just glance at the integral twice, when you need each one, rather than remember both of them. Also, glancing at the integrand can jog your memory of what term is next.

The polynomial above is a good example to work with, because it is a mid-size expression. Depending on the size of the expression and your own ability (as aided by the techniques in this chapter), you may be able to hold the whole integrated expression in your memory, and visualize each term as needed. If it seems to be larger than you know how to handle at that point, it may be best to write it down. You can then glance at it as you do the calculation.

To help calculate this, you can utilize a tool called *subimages*. Using subimages is kind of a cross between windowing and chunking. Imagine you have two cards floating in front of you, stacked over each other. On one of them you will keep a running total, and on the other you will put one of the terms in the expression above. You visualize one of them as having the number 0 on it:

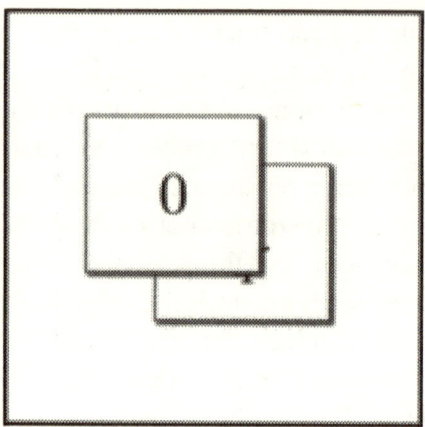

In your mind's eye, bring the bottom one on top, so that you can see it better:

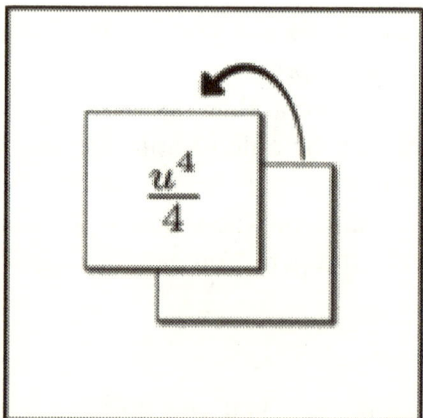

That second one has the first term in the antiderivative. Plug in 10 for u there (the value of the higher limit of integration), and evaluate it.

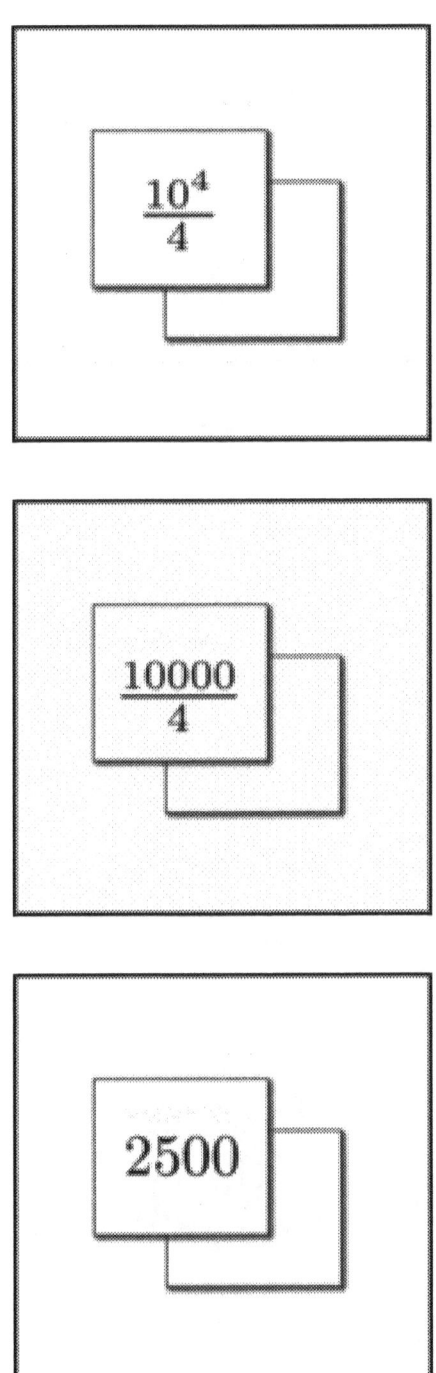

The other card is a running total, which of course started at zero. Now that you have the value of the first term, switch back to the running total card and add it in:

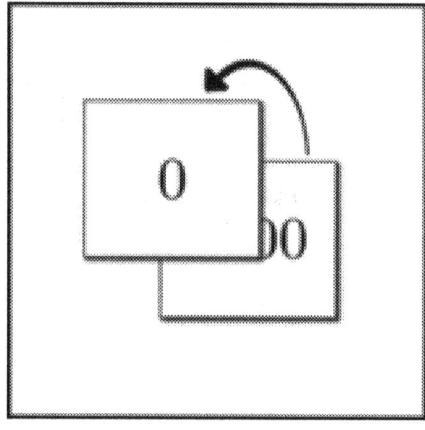

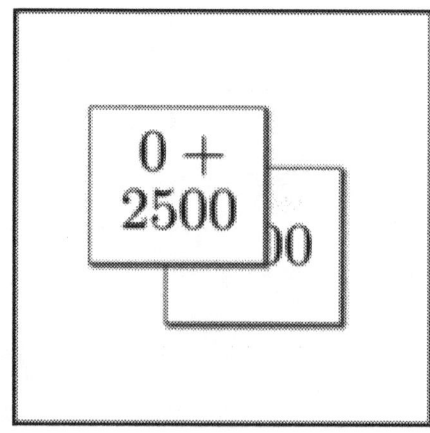

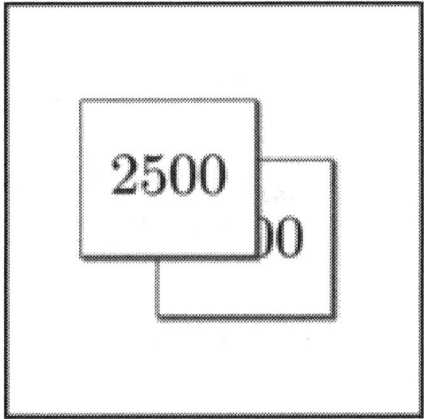

Then switch back to the other card, which now shows the next term in the series.

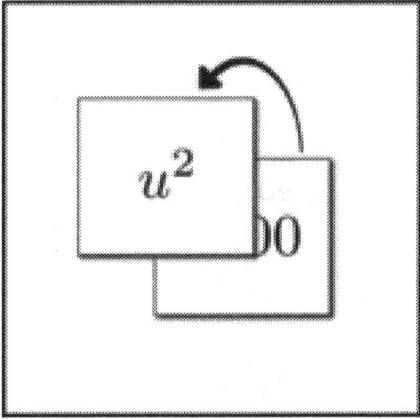

Again, plug in for $u = 10$, evaluate it, and add it into the running total.

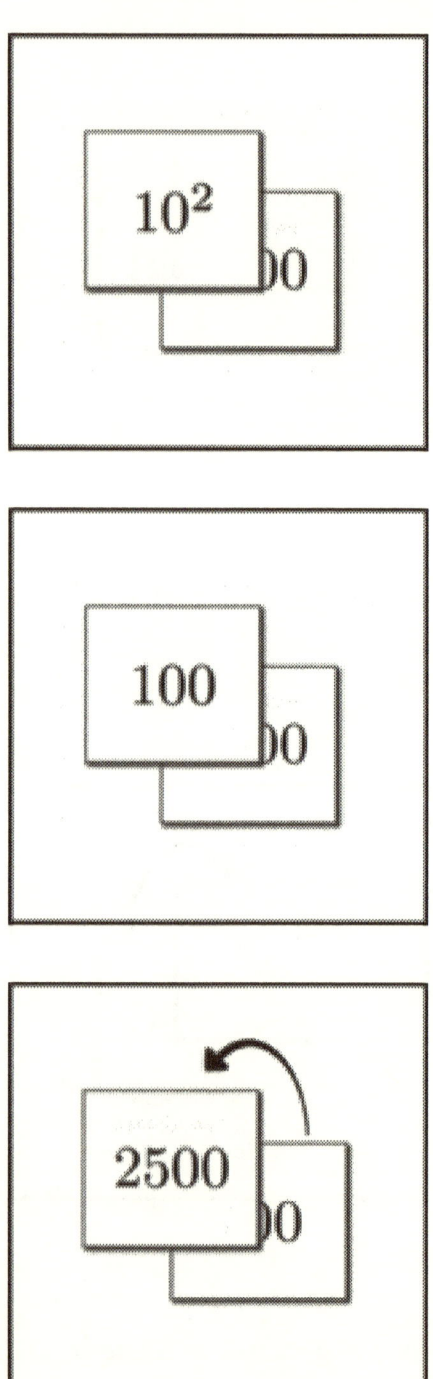

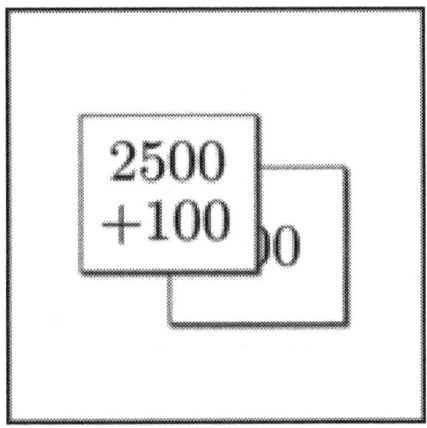

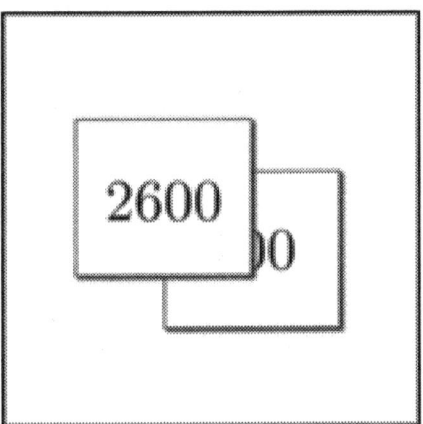

Repeat this process for each of the terms in the expression (in this example, there is only one left, -3u). That done, you run through them again. This time, you (a) use the other limit of integration, and (b) negate each term, multiplying it by -1. So after running through the expression the first time, you would visualize

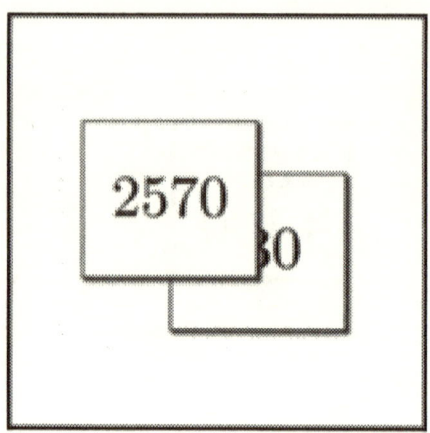

The top subimage, 2570, is the running total. That's the value of $\frac{1}{4}u^4 + u^2 - 3u$ at u = 10. The subimage below it, which you can't see well here, contains -30 (the value of the last term, -3u, at u = 10). At this point, we are ready to run through the terms again. Start by putting $-\frac{u^4}{4}$ in the subimage for the term, and evaluating it at u = 2:

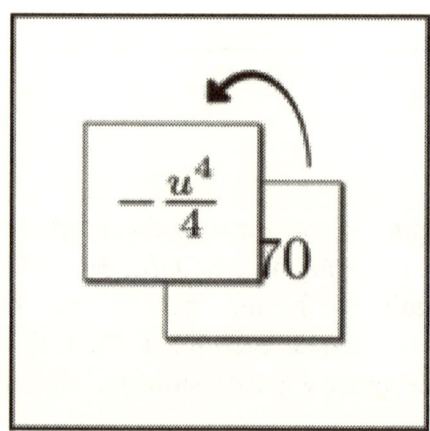

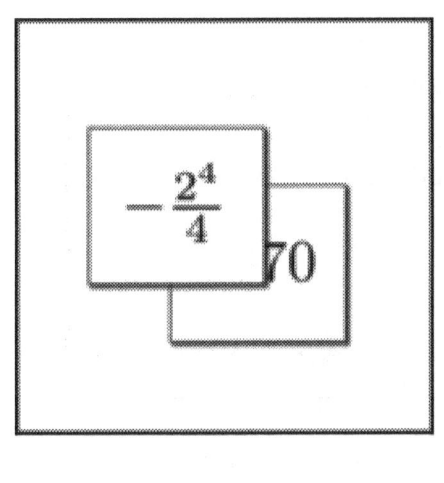

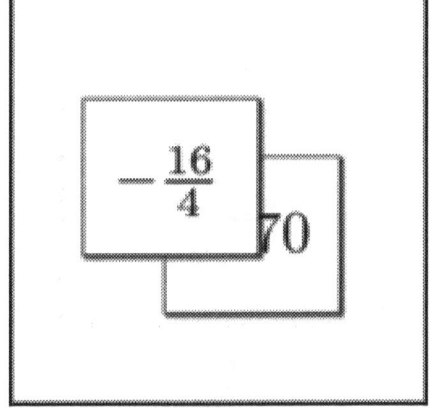

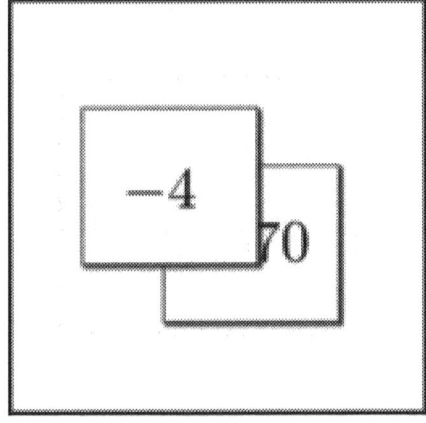

Do you understand why it is $-\frac{u^4}{4}$ instead of $+\frac{u^4}{4}$? It is, of course, just because $\int_a^b f(x)\, dx = F(b) - F(a)$, and we are now evaluating with the lower limit of integration.

Next, you add the value from this term into the subtotal, just as above.

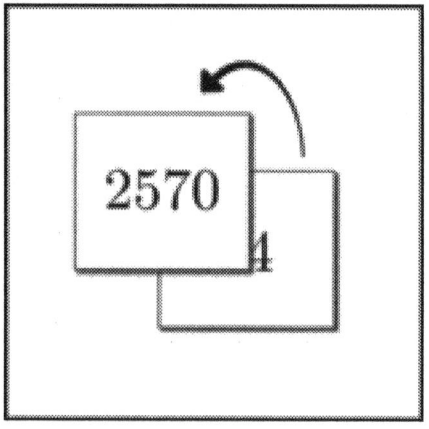

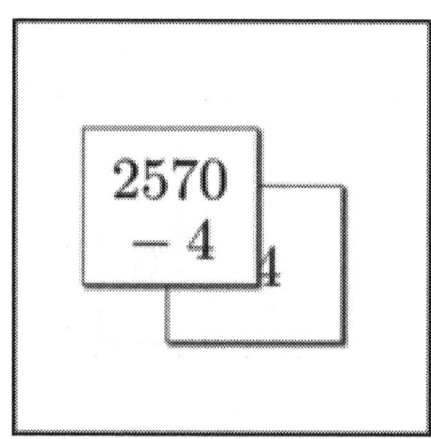

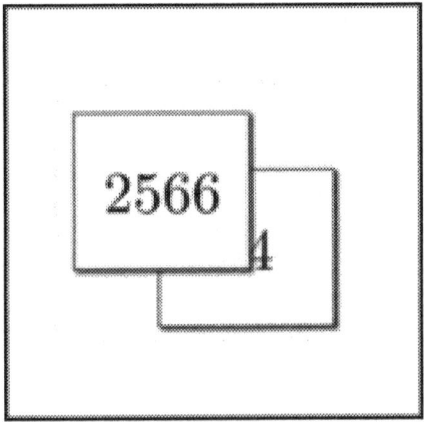

Continue this with the other terms. You finally will end up visualizing

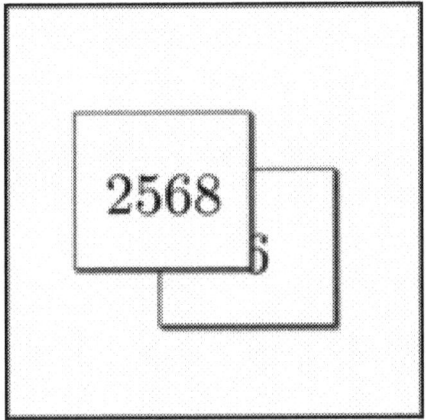

2568 is the final total.

Calculating From The Inside Out

The second principle that will help you calculate a complex expression mentally is to calculate from the "inside out". In other words, you calculate in the way that most rapidly reduces the complexity of the expression. Mastering this will help you do very complicated arithmetic almost as easily as with the simple cases.

The basic idea of this is that if you are evaluating some expression numerically - meaning, you have actual numbers to plug in for all the variables and now want to simplify the expression to a single number - you can choose to evaluate parts of the expression in a certain order that maximally simplifies the expression as you go along. At each point, ask yourself, "What part of this expression can I reduce next that will result in the simplest possible result?" Often, if you have an expression with many nested expressions, the best path is to calculate the most deeply nested ones first. This results in a simpler expression, which may also have nested elements; you then repeat the process again, only it's easier this time, since the expression has fewer symbols to keep track of.

As an example, consider using the quadratic formula to solve equations of the form $ax^2+bx+c = 0$. Say you want the solutions of the equation $2.2x^2 + 4x - 0.3 = 0$. Factoring this, mentally or not, would be a lot of work, so it's easiest to just calculate the solution directly with $x = \frac{1}{2a}\left(-b \pm \sqrt{b^2 - 4ac}\right)$.

The rule of thumb to use to decide what part of the expression to reduce next at any given stage is: *when deciding which part of an expression to reduce numerically, choose the part of that will result in the expression that has the fewest symbols.* In other words, you don't have to evaluate the the expression left-to-right; you choose to evaluate that part of the expression, anywhere within it, that will reduce it the most. When you are reducing a large, complex expression numerically, following this rule makes the process far easier and faster. There are some subtleties to the process, including in combination with windowing, which we will discuss now.

Look at the quadratic formula carefully:

$$x = \frac{1}{2a}\left(-b \pm \sqrt{b^2 - 4ac}\right)$$

Which symbols in the right-hand side can you reduce by plugging in numbers, that will give you an expression that is simpler (in terms of having fewer symbols)? By "simpler" here, we basically mean "easier to work with mentally". If you reduce the $\frac{1}{2a}$ to (approximately) 0.23, you end up with $0.23\left(-b \pm \sqrt{b^2 - 4ac}\right)$. That's probably a little easier to visualize, but not much. What if you plug in for the -b term? Well, you end up with

$$\frac{1}{2a}\left(-4 \pm \sqrt{b^2 - 4ac}\right),$$

which is about the same complexity as the original expression.

The other option is to calculate the discriminant. Depending on your skill and how "nice" the numbers are, you may mentally calculate $b^2 - 4ac$ in one step. In this case, that would leave you visualizing $\frac{1}{2a}\left(-b \pm \sqrt{18.64}\right)$. If that is not doable, you can apply the inside-out strategy to the discriminant itself. The inside-out strategy is recursive! Temporarily forget about the larger expression, and just calculate the value of $\sqrt{b^2 - 4ac}$. (This is where windowing comes in.) What part of this expression can you reduce by plugging in numbers, which will simplify the expression the most? The answer is probably the $4ac$ term. Reducing that term gives you the expression $\sqrt{b^2 + 2.64}$. Visualizing that, you can next substitute in for the b^2 and simplify to get $\sqrt{18.64}$. Since this is not a perfect square, we can choose to approximate a value of 4.4. (The actual value is approximately 4.32.) At this point, you can visualize the original, full expression, which now looks like $\frac{1}{2a}\left(-b \pm 4.4\right)$. Much easier to visualize than $\frac{1}{2a}\left(-b \pm \sqrt{b^2 - 4ac}\right)$, isn't it? By properly using the "inside-out" strategy, you can rapidly whittle very complex expressions down to more manageable sizes. Since this expression has a plus-or-minus symbol, you will at some point have to split up your image into two parts, one for each solution[1]. Regardless,

---

[1] The chapter of *Inner Algebra* entitled "Plurality" teaches some specific techniques for working with equations with multiple solutions. It is based on subimages, similar to what was done above.

you can continue reducing it from the inside out in a similar way.

Chapter 5

Closing Thoughts

This book is rather different from most math textbooks. While most math books are about some set of math topics, this book is mainly about *you*. It's about what happens inside of you when you do math. It's about the effects of what you do inside have on how easy or hard it is for you to do things in the world. Most of all, it is about how you can take ownership of your own internal processes, gaining greater freedom of expression.

Regardless of whether you use mathematics heavily in your life's work, please take what you have learned here and consider how you may apply it in those and related areas. If you have difficulty in a certain situation, look at yourself first, and ask if there is something in your attitude or perception that can open to a different, more successful approach.

Some of the happiest moments of my life have been when I was working some expression out in my mind's eye. My hope is that through this book, you may find a bit of that joy for yourself. Thank you for reading this book.

Index

About the Author

Aaron Maxwell has a B.S. in physics from the University of Texas at Dallas. He has had graduate-level training in biophysics and neuroscience, and has worked as a software developer. He chose these roles in part because of the fascinating ways that mathematics are applied in them. He is the founder and owner of Hilomath (http://hilomath.com), a math education company. Currently he lives in San Francisco, California, USA. He can be contacted by email at amax@hilomath.com .

www.ingramcontent.com/pod-product-compliance
Lightning Source LLC
Chambersburg PA
CBHW021931170526
45157CB00005B/2273